Rocky Roots

GEOLOGY AND STONE CONSTRUCTION IN DOWNTOWN ST. PAUL

Second Edition

Sister Joan Kain
and
Paul D. Nelson

Ramsey County Historical Society
St. Paul, Minnesota

Production Credits

Copyediting: John M. Lindley, John M. Lindley & Associates
Text Design: Wendy Holdman, BookMobile Design and Publishing Services
Composition: BookMobile Design and Publishing Services
Cartography: Patti Isaacs, Parrot Graphics
Cover Design: Kyle Hunter, BookMobile Design and Publishing Services
Photography: Maureen McGinn and Paul D. Nelson
Production Coordination: Connie Kuhnz, BookMobile Design and Publishing Services
Production Management: John M. Lindley, John M. Lindley & Associates
Printing and Binding: BookMobile

Copyright © 1978 and 2008 by
Ramsey County Historical Society

The first edition of this book was published by the Ramsey County Historical Society in 1978 and had the title *Rocky Roots: Three Geology Walking Tours of Downtown St. Paul.*

Published by
Ramsey County Historical Society
323 Landmark Center
75 West Fifth Street
St. Paul, MN 55102
(651) 222-0701
www.rchs.com

All rights reserved. No part of this book may be reproduced in any form or by any electronic or mechanical means, including information storage and retrieval systems, without permission in writing from the publisher, except by a reviewer, who may quote brief passages in review. Any members of educational institutions wishing to photocopy part or all of the work for classroom use, or publishers who would like to obtain permission to include the work in an anthology, should send their inquiries to the Ramsey County Historical Society, 323 Landmark Center, 75 West Fifth Street, St. Paul, MN 55102.

Printed in the United States of America.

ISBN-13: 978-0-934294-68-3

Cover photos: The background photo is Laredo Chiaro marble from the St. Paul City Hall/Ramsey County Courthouse (8). The individual photos (right to left; the numbers in parentheses indicate the building identification number in this book) across the front and back covers are the Cathedral of St. Paul (16); the St. Paul City Hall/Ramsey County Courthouse (8); the Warren E. Burger Federal Courts Building (34); the Minnesota State Capitol (18); the Cathedral of St. Paul (16); the Railroad and Bank Building (39); the First National Bank (32); the Minnesota State Capitol (18); the First National Bank (32); the James J. Hill Library and St. Paul Central Library (5); the Qwest Buildings (6); the Landmark Center (4); and the Qwest Buildings (6).

CONTENTS

Preface to the Second Edition	vi
Rocky Beginnings	1
Architectural History of the City	2
Building Profiles	5
Master Map of Downtown St. Paul	20–21
Building Profiles	22
A Brief Geological History of Minnesota	35
The Quarries	36
Geological Terms	40

Carved detail in Mankato-Kasota stone, Qwest Buildings.

ACKNOWLEDGMENTS

Paul D. Nelson thanks Sister Joan Kain for her early guidance in this second edition; Steve Erickson of the Minnesota Geological Society, Jeff Thole of the Macalester College Geology Department, and Todd Olson of Cold Spring Granite Company for invaluable assistance in getting things right; Debbie Erickson at the St. Paul City Hall/Ramsey County Courthouse for her helpful answers to all my questions about that St. Paul landmark; Maureen McGinn for her photography and good judgment; Priscilla Farnham and John M. Lindley for getting this project off the ground. He also credits these excellent works: Larry Millett, *AIA Guide to the Twin Cities: The Essential Source on the Architecture of Minneapolis and St. Paul* (Minnesota Historical Society Press, 2007) and Jeffrey A. Hess and Paul Clifford Larson, *St. Paul Architecture: A History* (University of Minnesota Press, 2006).

The Ramsey County Historical Society thanks the following individuals and organizations for their generous assistance in making this new edition of *Rocky Roots* possible.

Tom Bergin Jr., Historic Stone Company
Laurence and Caroline Best
Rebecca Ganzel
Jennifer H. Gross and Jerry Lefevre
Frank Langer, Historic Stone Company
Julie Neville
Christine Podas-Larson
Vetter Stone Company, Kasota, Minnesota

The Ramsey County Historical Society joins the following persons in honoring Sister Joan Kain whose dedication to teaching geology inspired her to write the first edition of Rocky Roots.

In Honor of Sister Joan Kain
Susan Nemitz and John Curry
Wendy Nemitz and Scott Kramer
Elle Kokkinos
Margaret and Thomas Swifka
Kristi and Larry Waite

The Ramsey County Historical Society also joins with the following individuals and organizations in remembering five persons who gave many years of support to the Society and local history.

In memory of Robert Farnham
Malcolm and Patricia McDonald

In memory of Henry Blodgett, Charlotte Drake, Virginia Brainerd Kunz, and Susan Hill Lindley
Richard and Nancy Nicholson
Nicholson Family Foundation

PREFACE TO THE SECOND EDITION

This *Rocky Roots* brings up to date and expands the original *Rocky Roots*, published in 1978. The inspiration remains that of its first author, Sister Joan Kain, and many of her words have been preserved too. The 1978 book documented and celebrated the use of stone, mostly structural, in buildings in downtown St. Paul. There was a bit of quirky genius in this. Stone is all around us; we see it, yet it registers lightly if at all in our daily consciousness. Sister Joan and her collaborators, Jerry Backlund and Dennis Brinkman, undertook to help us all see that in building stone there are stories, both geological and human. The book has enjoyed steady demand for thirty years. In that time, of course, much has changed—old buildings destroyed, new ones built. Hence the opportunity and need for a new edition of *Rocky Roots*.

This project presented some easy decisions, such as deleting from the text all mention of the buildings that no longer exist, and some harder ones: how much of the original text and organization to keep, whether to retain all of the other buildings discussed in the original, even if some observers consider them undistinguished, and which new buildings to add.

In choosing buildings to highlight we used two general criteria: quantity and quality. To be given more than passing mention in this book, a building must have a lot of stone: quantity. It must also have an interesting quality—historical significance, beauty, or distinctiveness. Hence the Hamm Building, is excluded: though beautiful, it has hardly any stone. The Transportation Building, a great mass of stone sheathing, gets scant mention for one reason: its use of stone is pedestrian. Some stone buildings are not mentioned just because there is not much to say about them.

The original black and white photographs of the first edition all had to be replaced, requiring a whole new set of photographs and decisions.

The 1978 book dealt very little with interior stone, but in doing research (that is, poking our heads in the buildings) we were reminded that three St. Paul buildings in particular, the state Capitol, the St. Paul City Hall/Ramsey County Courthouse, and the Cathedral of St. Paul, offer a stupendous array of decorative stone, mainly marble, from all over the world. So we added interior stone to the book.

The original *Rocky Roots* was presented as "three walking

Decorative bat ornamentation, Landmark Center.

tours." We came to feel (as we walked) that now the buildings with noteworthy stone were too scattered to permit of a walking-tour organization that made walking sense. So we gave that up in favor of simply highlighting and mapping the buildings; the reader must decide how to visit them, without editorial guidance. We provide the *Roots*, the routes are up to you.

With all the changes (and there are others), we believe Sister Joan's inspiration persists intact. She and we invite readers to look at the built world around them with new eyes. The stone most of us are accustomed barely to notice is not just rock; every variety has a history, a composition, a provenance, and sometimes a vogue. There is pleasure in finding something new in the familiar.

ROCKY BEGINNINGS

Downtown St. Paul stands on a rather thin shelf of limestone. Beneath that shelf, there is soft St. Peter sandstone. These formations are remnants of an inland sea that covered this part of the world 450 to 500 million years ago. Underneath these sedimentary rocks lies a trough of basaltic lava created by volcanic action a billion years past.

The Mississippi gorge that separates downtown St. Paul from the West Side traces the path of River Warren Falls, which 12,000 or so years ago thundered magnificently where bridges cross the river today. The floods that once filled the gorge from bluff to bluff surged across a crumbling limestone ledge, then fell, from heights as great as two hundred feet. The cascading water eroded the soft sandstone beneath the limestone, leaving the exposed edges without foundation. As these broke off, the falls marched upstream toward what is now Fort Snelling, eventually dwindling to nothing up the Minnesota River valley. A remnant continued up the Mississippi valley to what became St. Anthony Falls in downtown Minneapolis. The human hand stopped it there.

A visitor to downtown St. Paul can easily see the limestone shelf and sandstone foundation. Step out a few yards on the east side of the Wabasha Bridge, then turn and look back toward the city, then move your eyes down and to the right. The limestone is just a few feet thick in this location; the sandstone beneath it is so soft that birds excavate it for their nests.

Platteville limestone above Glenwood shale and St. Peter sandstone. Dayton's Bluff, St. Paul.

St. Paul's limestone foundation could not supply the high quality building stone that a growing and sometimes prideful metropolis required. Hence the stone that one finds downtown is mostly from quarries outside the St. Paul area. The dimension stone—granite, sandstone, dolostone, limestone, gneiss—mostly comes from the Midwest. The interior ornamental stone, chiefly marble, comes from all over the world.

Every stone carries an ancient story, and every building—a human creation of imagination and toil—another story or many. This book aims to bring to light a few fragments of these stories, where the stones of the earth and human mind and hand come together in architecture and craft.

ARCHITECTURAL HISTORY OF THE CITY

Downtown St. Paul clusters on a bluff above a sweeping curve of the Mississippi River. The city was founded, to use a term too grand for a squalid and desperate beginning, in 1840 when squatters evicted from the Fort Snelling military reservation moved down the river to settle at "White Rocks," or *Im-in-i-ja Ska*, as the site of St. Paul was called in a Dakota

The Chapel of St. Paul. Photo courtesy of the Minnesota Historical Society.

language. Its first English name, Pig's Eye Landing, probably suited the muddy hamlet well. But by 1848 the village had become St. Paul, taking the name from its small, log-built Roman Catholic chapel.

Located at the northern limits of practical navigation of the Mississippi River, the frontier town welcomed settlers and goods brought upstream by the glorious new technology of the time, steamboats. Within forty years a more powerful technology, railroads, turned the river burg into a boomtown, and then into a city.

The structures of the first few decades were mostly frame, hurriedly built and easily replaced (or burned). The first building stone used was the Platteville limestone lying

Downtown in 1871. Photo courtesy of the Minnesota Historical Society.

conveniently under foot. Only a few old limestone buildings remain, the most prominent being Assumption School (1864) and Assumption Church (1874). Brick, cheaper and easier to use, replaced the Platteville stone as the city grew. As rail lines extended their reach, Mankato-Kasota stone came in from southern Minnesota (First Baptist Church, 1875), red sandstone arrived from Wisconsin and Michigan (Central Presbyterian Church, 1889), and granite from St. Cloud (Pioneer and Endicott Buildings, 1889).

Many buildings put up from the 1890s to the 1920s, the era of St. Paul's glory, reflect the influence of Chicago's 1893

Third Street (now Kellogg Boulevard) in 1862. Photo courtesy of the Minnesota Historical Society.

World's Columbian Exposition. These include St. Paul's signature monuments, Cass Gilbert's State Capitol (1906) and Emmanuel Masqueray's Cathedral of St. Paul (1918). The Union Depot (1923) and Judicial Center (formerly Minnesota Historical Society, 1917), with their colonnaded entrances, represent a more restrained neoclassical style.

The flowering of Art Deco in the 1930s, best represented by the Jemne Building (1931) and the St. Paul City Hall/Ramsey County Courthouse (1932), preceded more than half a century of destruction of fine old buildings, and pedestrian (or worse) design of new ones. So far as the use of stone is concerned, the Minnesota History Center (1991) and new state office buildings near the Capitol mark welcome and hopeful achievements.

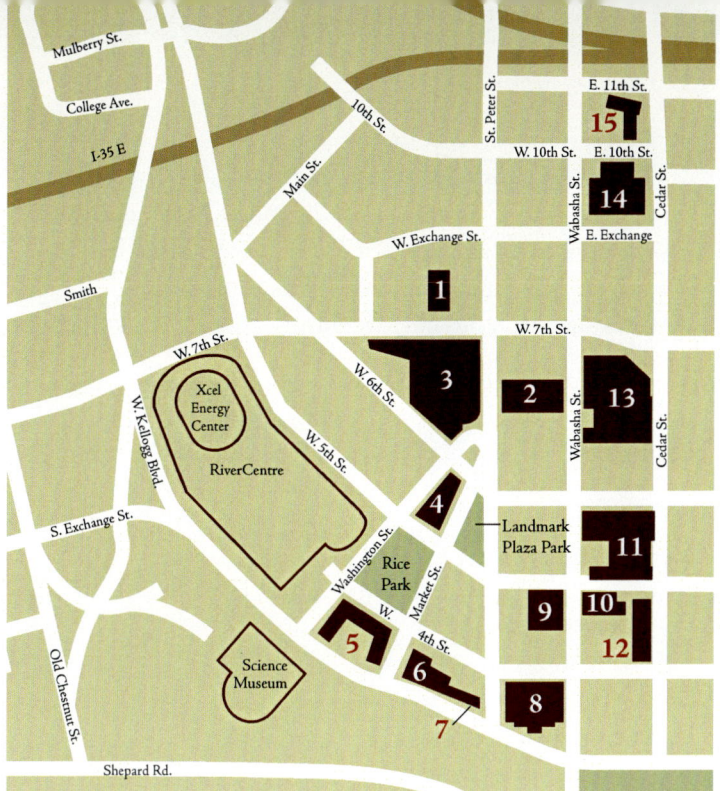

1. Assumption Church
2. Coney Island
3. Travelers
4. Landmark Center
5. James J. Hill Library & St. Paul Central Library
6. Qwest Buildings
7. Jemne Building
8. St. Paul City Hall & Ramsey County Courthouse
9. St. Paul Building
10. Ecolab Building
11. Ecolab Center
12. Pioneer Press Building
13. Wells Fargo Place
14. McNally-Smith College of Music
15. St. Paul Department of Public Health Building

1

Assumption Church   < 1873 >
51 West Ninth Street

The church is St. Paul's oldest, and the most prominent symbol of the city's German Catholic heritage. Though St. Paul has cultivated an Irish image, for 150 years its predominant ethnic group has been German. The parish dates from 1854 and this church from 1873 Rare among city buildings, it is made of the

Assumption Church apse.

< 5 >

Platteville limestone that underlies the city itself. The style is Romanesque, the design based on the *Ludwigskirche* of Munich. A visit to the cool, serene interior is recommended.

The former Assumption School next door is downtown's second-oldest building, 1864, made of the same Platteville limestone, probably quarried nearby.

2

Coney Island < 1858 >
448 St. Peter Street

The smaller of the side-by-side buildings is downtown's oldest. It served as the state arsenal from 1865 to 1880. The stone is Platteville limestone.

Platteville limestone and brick, Coney Island.

3

Travelers < 1991 >
385 Washington Street

There are two buildings here, separated by Fifth Street, the older one built in 1961, the next one thirty years later. Both feature facing of St. Cloud granite and details of Mankato-Kasota dolostone. The interplay of these two local stones can best be seen walking north on St. Peter Street toward Mickey's Diner. The chubby round tower of the newer building (designed by St. Paul native William Pederson) appears to make reference both to the Landmark Center and the apse of nearby Assumption Church. The lobby mural shows St. Paul from Fort Snelling to downtown in the 1850s.

Travelers headquarters; Mankato-Kasota stone below, St. Cloud granite above.

Landmark Center   < 1906 >
75 West Fifth Street

This was St. Paul's main post office for nearly thirty years and the federal courthouse from 1906 to 1969; Alvin Karpis and the Ma Barker gang (those who survived the shootout, that is) received their long prison sentences here. Its turrets, corner tournels and stair towers, gables, and steeply pitched tiled roof are reminiscent of the *chateaux* of the Loire Valley in France. The massive base of the building with its rounded arches, carved details, and commanding square tower has the characteristics of the Romanesque style of architecture. The stone is Sauk Rapids pink granite, which does in fact look a bit pink up close. Reddish-burgundy marble wainscoting, which may have come from northwestern Vermont, graces the main hall and the stairway.

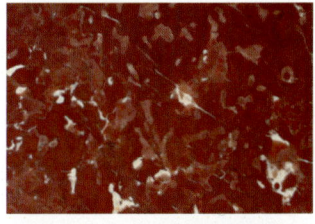

Only a determined and last-ditch citizens' campaign, led by Elizabeth Musser, Georgia DeCoster, and others, saved this charming and irreplaceable building from the wrecking ball. The succeeding renovation, completed in 1979, restored the airy central atrium and returned the place to its original beauty.

Marble wainscoting and a decorative granite medallion, Landmark Center.

5

James J. Hill Library and St. Paul Central Library < 1916 >
80–90 West Fourth Street

Pink Tennessee marble was used in St. Paul's classical library building, a gift, in part, of James J. Hill, who established a separate and private business reference library in the east wing. Hill, a man who insisted on having his way, chose Electus D. Litchfield of New York as architect. He produced an elegant, restrained Florentine "palace" near the site of St. Paul's old farmers' market. The marble looks white from a distance, but distinctively pink (especially when wet) up close. The interior walls are Winona travertine. The beautiful center stairway is a distinguishing feature of the public library part of the building.

The main reading rooms of both libraries are very fine. The public library was restored and remodeled in 2002.

Library cornice.

Pink Tennessee marble.

6

Qwest Buildings < 1935, 1968, 1976 >
70 West Fourth Street

There are three buildings here, side by side by side. From west to east, they were built in 1935, 1968, and 1976. The purple cast of the newest, the brick one, blends with the maroon colors in the Morton gneiss, or Rainbow Stone, that forms the base of the two earlier Art Deco structures (and of the former West Publishing building across the street). The Rainbow Stone quarried near Morton, Minnesota, is one of the oldest exposed rocks on the earth. To builders it is colorful, distinctive, available, and easily worked. The dark base and trim of the earlier telephone building forms an interesting contrast to the Mankato-Kasota stone of the upper floors. It is well worth raising one's eyes to the Maya-like carvings above the windows of the 1935 building. The 1968 building next door is mostly Kasota stone too, with its blocks laid vertically rather than horizontally.

Rainbow Stone frames a Qwest Buildings entrance.

Exterior detail in Mankato-Kasota stone, Qwest Buildings.

7

Jemne Building < 1931 >
305 St. Peter Street

The walls of this Art Deco structure are Mankato-Kasota stone panels in a buff color with a travertine (that is, marked and pitted) texture. They form the angled curve that gives the building its distinctive geometric styling. The architect, Marcus Jemne, shaped the building to meet the challenge of an oddly shaped site and still retain a view of the Mississippi. The base is polished black stone

The Jemne Building— Mankato-Kasota stone.

often confused with granite, the same Veined Ebony from Mellen, Wisconsin, used for the base of the courthouse across the street.

"Is there a lovelier small building in the Twin Cities than this jewel . . . ?" wrote architectural historian Larry Millett. "[T]he building has a delicacy rarely seen in institutional architecture yet also holds down its corner site with convincing authority."

8

St. Paul City Hall/Ramsey County Courthouse < 1932 >
15 West Kellogg Boulevard

Indiana limestone panels alternate with dark spandrels and narrow framed windows to emphasize the vertical lines of St. Paul's Art Deco monument to city and county government. The building's base is Mellen Veined Ebony, a variety of igneous rock called olivine gabbro.

It is best to enter the building from the north, greeted by the relief sculpture of an idealized St. Paul street scene above the doors. Inside, the main hall has no local rival for dramatic use of marble—thirty vertical feet of the nearly black Bleu Belge (from Belgium, a stone now exhausted), with a polished gold ceiling, leading to Carl Milles's towering white Mexican onyx (from Baja California) statue, "Vision of Peace." The names carved into the walls are those of Ramsey County war dead from World War I forward.

The rest of the main floor walls are Laredo Chiaro from Italy, and the floors alternating squares of French Champville and Hauteville. Second and third floor walls repeat the Bleu Belge/Laredo Chiaro pattern, while the floor changes to Roman travertine. There is a fine multi-colored terrazzo map of Ramsey County north of the elevators on the second floor. Tucked away in the mayor's conference room are baseboards of Red Lepanto from Liguria, near Genoa.

Bleu Belge marble panel, Memorial Hall.

Often the best places to appreciate the color and patterns of the marbles are in the drinking fountain recesses: Westfield Green from Massachusetts, for example, on floors four, six, seven, and twelve; Verde Issorie on five and seventeen. This last is not original, but added in the 1990s renovation. Mount Nebo Golden Travis marble from Utah graces the lobbies of floors eight through eleven. Dark green Tinos from Greece, (present also in the Cathedral) may be found in office entrances on four, six, seven, nine, and twelve. Breche Du Nord from Belgium appears in courtrooms on the fourteenth and fifteenth floors. In the basement, completely remodeled in the early 1990s, the wainscoting is Botticino from Brescia, Lombardy, the bases black and white Pyrenees from Spain.

The Vision of Peace, Memorial Hall. Photo by George Heinrich.

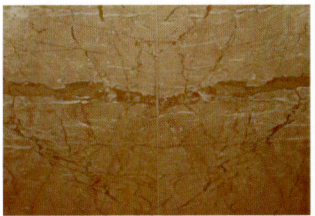

Laredo Chiaro marble.

There are fourteen varieties of marble on display in what Larry Millett calls "a masterpiece of American art deco."

The City Hall Annex, across Fourth Street, has some nice touches—Indiana limestone and big panels of Mellen Veined Ebony that mirror City Hall—plus small rectangles of Vermont Verde Antiqua (a serpentinite) beneath the second floor windows.

Bas relief street scene in Indiana limestone above the north (Fourth Street) entrance. The sculptor was Lee Lawrie.

9

St. Paul Building < 1889 >
6 West Fifth Street

The St. Paul Building.

The former Germania Bank building is an excellent example of an 1880s commercial structure. The design shows Henry Hobson Richardson's influence in the use of brownstone (Jacobsville sandstone from the Upper Peninsula of Michigan) columns and arches, in the massed window treatment, and in the building's decorative patterns. The sandstone looks structural, but it is not. Architect J. Walter Stevens also designed several fine brick commercial and warehouse buildings in Lowertown, still standing.

10

Ecolab Building < 1930 >
360 Wabasha Street

Ecolab Building.

This Art Deco building began its life as the Northern States Power building, as reflected in the bronze relief over the entrance, Light, Heat, and Power. The Rainbow Stone base is particularly fine. If you take a few minutes to examine it closely, you can see the tumultuous forces of the ancient earth—intense heat and pressure—captured, frozen in the panels. The blobs and swirls were molten stone churning deep, and incomprehensibly slowly, miles beneath the surface. Above, Makato-Kasota dolostone; the combination resembles that of the 1935 Qwest building.

< 12 >

Across Wabasha, in what remains of the façade of the old Lowry Lounge, are two delightful relief panels in Mankato-Kasota stone portraying Pan and a nymph, both looking pleased with themselves.

Pan playing his pipe, façade of the former Lowry Lounge on Wabasha Street.

11

Ecolab Center < 1968 >
370 North Wabasha Street

Originally the Osborn Building, it is mostly stainless steel and glass, but there is plenty of black stone, too, in the pillars around the base, on the benches and low walls of the plazas. Larry Millett calls the plaza behind the building, invisible from the street (and off-limits to the public), the best downtown, "with sinuous black granite benches" and an abstract sculpture "that serves as a perfect foil to the resolutely rectilinear building." To quibble, there is no such thing as black granite, though the phrase is often used. This stone is a gabbro; if you examine it closely, you will see that it lacks the variety of crystals characteristic of granite.

Mellen Veined Ebony, Ecolab Center.

12

Pioneer Press Building < 1955 >
345 Cedar Street

The black base is not stone, but a man-made composition that resembles stone from a distance. The walls above are Mankato-Kasota. This used to be the headquarters of Minnesota Mutual Life Insurance Company, now Securian, whose new headquarters is just as routine as this one.

13

Wells Fargo Place < 1987 >
Wabasha Street
and Seventh Place

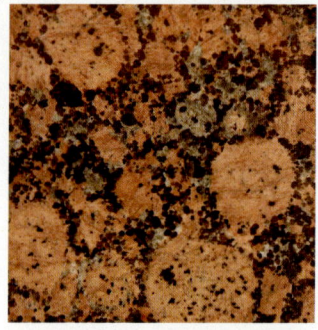

Carmen Red granite, Wells Fargo Center. Photo by Jeff Thole.

This building began life as the Minnesota World Trade Center, a grand concept—St. Paul, a hub of international commerce—forgotten long ago. It still makes a strong, tall, and very maroon statement with its exterior walls of Carmen Red granite, a stone quarried in Finland. Up close, one can see the striking contrast between the polished and unpolished versions of the stone.

14

McNally-Smith College of Music < 1964 >
30 East Tenth Street

The former Minnesota Arts and Science Center is not beautiful, but the three-stone composition—Indiana limestone, Iridian (from Isle, Minnesota), and Charcoal Black granite—achieves a handsome dignity.

15

St. Paul Department of Public Health Building < 1957 >
555 Cedar Street

This low, L-shaped building attracts little notice, but it has some style—a later *Moderne* severity and very large blocks of Mankato-Kasota stone in a warm hue.

Public Health Building, Mankato-Kasota stone.

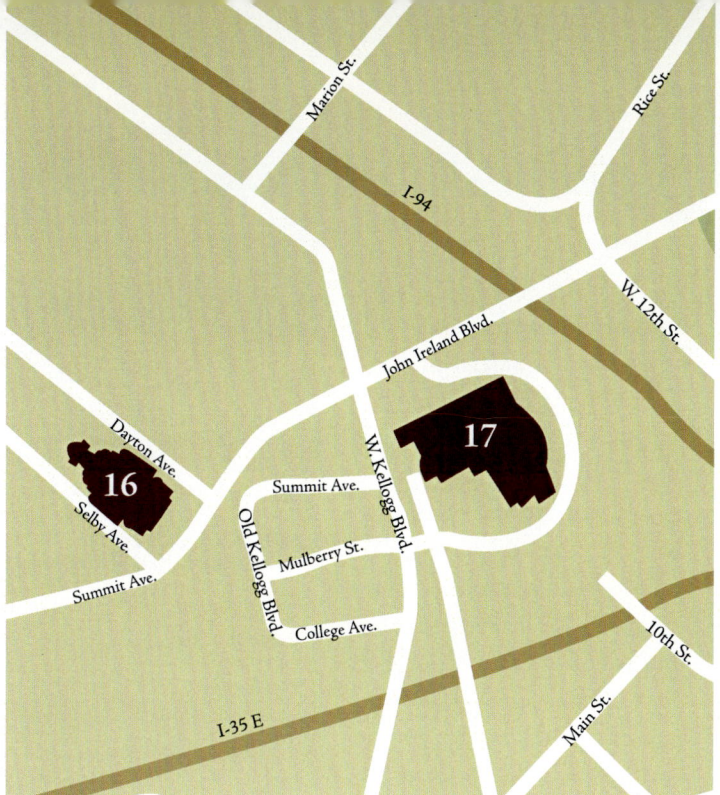

16. Cathedral of St. Paul   17. Minnesota History Center

16

Cathedral of St. Paul < 1915 >

225 Summit Avenue

The Church of St. Paul still watches over the city—not as a log chapel at the edge of a straggling settlement, but as a domed and towering monument to God's glory rising above a busy metropolis. Rockville

Cathedral of Saint Paul. Photo by Doug Ohman.

granite, Mankato-Kasota dolostone, and the recently restored copper dome form this baroque structure designed by the Frenchman Emmanuel Masqueray.

The severe exterior belies a glorious, awe-inspiring, and colorful interior. The best way to enter is up the main staircase and through the doors that open over the city skyline. The first color one senses is green, the Verde Antico marble,

< 15 >

from Thessaly, Greece, of the narthex, the long transverse hall that separates the exterior doors from the interior proper.

The main seating area, beneath the soaring dome, features a six-foot wainscoting of Tavernelle marble from France, and above it Winona travertine. Most of the marble in the building is in the sanctuary and in the eleven chapels (the twelfth recess is the Baptistry) arranged around the great nave and choir.

The centerpiece of the sanctuary is the main altar, cut from Botticino marble, with a front panel of Rojo Alicante from Spain and borders of Verona Red and Giallo Fantastico, both from Italy. The black and gold columns are Portora. The stairs leading up to the altar are Istrian, the floor Alpine green with insets of Levanto Red, Swiss Cipollino, and Sylvan Green. The eight columns between the arches are Botticino, and above them ovals of Verona Red and Greek Tinos.

The sanctuary is merely the most spectacular expression of marble craft and color in the building. Every one of the chapels has its unique qualities. Starting at the back of the nave, and proceeding to the left, the Chapel of the Blessed Virgin Mary features a remarkable panel of Vert Clair behind the statue. At the front, to the left of the Sanctuary, the Chapel of St. Peter employs a host of

Verde Antico marble, Cathedral narthex.

St. Peter Chapel. Photo by Doug Ohman.

Marble railing, St. Peter chapel.

marbles: from the famous quarries at Carrara: Sienna gold in the walls and columns, Sienna gray in the altar, brown marble in the tables, white marble in the floor with strips of Giallo Fantastico ("fantastic yellow") from Serravezza.

Steps to the right of this chapel lead to the Shrine of Nations, six small chapels arrayed in a semicircle, dedicated to ethnic groups that contributed to the building of St. Paul and its cathedral. The common elements are the walls, of Botticino, and the statues of the various saints, all Trani marble quarried and carved in Italy. Four of the chapels feature marble from the corresponding country: a panel behind the saint and in the front panel of the altar, polished, and unpolished in a rondelle on the floor. For St. Patrick (Irish), Connemara Green (from County Galway); for St. Boniface (German), German Formosa (near Nassau, in the Rhineland); for St. John the Baptist (French), Swiss Cipollino. For the Sts. Cyril and Methodius chapel (Slavic), the marble is St. Genevieve Golden Vein from Missouri.

The chapel of the Scared Heart, just north (to the right) of the Sanctuary, is a riot of ornamental stone: Laredo Chiaro from Italy, Numidian Cipollino and

Chapel of St. Anthony. Photo by Doug Ohman.

Breccia Violetta marble pillar in the Chapel of St. Joseph.

Greek Tinos marble, Cathedral vestibules.

Numidian Crimson from Algeria, Rojo Alicante from Spain, green Tinos from Greece, Portora black and light Botticino from Italy again, and Moroccan Red onyx.

In the Chapel of St. Joseph, back near the main entrance and opposite the Chapel of the Blessed Virgin, Breccia Violetta, Italian, for the columns; American Danbury Cream, from Vermont for the walls and altar; Tennessee pink in the floor; German rose and black Formosa for the panel behind the statue of the saint.

The three side vestibules are lined with Tinos, a rich green, from Greece.

17

Minnesota History Center < 1991 >

345 West Kellogg Boulevard

No building uses stone more pointedly than this one. The walls were built in alternating bands of Minnesota-quarried stone, Rockville White granite from Cold Spring and Winona travertine. The floors of the main concourses are laid in Minnesota granite and travertine with panels of the same bright, white Georgia marble used in the exterior walls of the state capitol.

Minnesota History Center.

Rockville white granite and Winona Travertine, History Center interior.

18. Minnesota State Capitol
19. Minnesota Judicial Center
20. State Office Building
21. Transportation Building
22. Veterans Service Building
23. Centennial Office Building
24. St. Paul National Guard Armory
25. Orville Freeman Building
26. State Laboratory Building
27. Harold Stassen Department of Revenue Building

18

Minnesota State Capitol ‹ 1906 ›

75 Rev. Dr. Martin Luther King Jr. Boulevard

Minnesota's serenely white State Capitol gives no hint of the lengthy legislative battle over the choice of building stone. Popular opinion and local interests wanted a Minnesota stone. Cass Gilbert, however, insisted on Georgia marble,

A 1907 postcard view of the Minnesota Capitol. Photo courtesy of the Minnesota Historical Society.

‹ 19 ›

MASTER MAP OF DOWNTOWN ST. PAUL

1. Assumption Church
2. Coney Island
3. Travelers
4. Landmark Center
5. James J. Hill Library & St. Paul Central Library
6. Qwest Buildings
7. Jemne Building
8. St. Paul City Hall & Ramsey County Courthouse
9. St. Paul Building
10. Ecolab Building
11. Ecolab Center
12. Pioneer Press Building
13. Wells Fargo Place
14. McNally-Smith College of Music
15. St. Paul Department of Public Health Building
16. Cathedral of St. Paul
17. Minnesota History Center
18. Minnesota State Capitol
19. Minnesota Judicial Center
20. State Office Building
21. Transportation Building
22. Veterans Service Building

23. Centennial Office Building
24. St. Paul National Guard Armory
25. Orville Freeman Building
26. State Laboratory Building
27. Harold Stassen Department of Revenue Building
28. Public Safety Building
29. Central Presbyterian Church
30. Golden Rule Building
31. Securian Center, 401 Building
32. First National Bank
33. Minnesota Building
34. Warren E. Burger Federal Courts Building
35. Pioneer Building
36. Endicott Building
37. Agribank
38. Brooks (McColl) Building
39. Railroad and Bank Building
40. Main Post Office
41. Union Depot
42. First Baptist Church

and contractors Walter and William Butler brought it in at a favorable price. Compromise ended the sometimes bitter discussions. Hinckley sandstone was chosen for the foundation piers of the building and dome; Winona limestone and local blue limestone were laid as foundation walls; granite from Baxter's quarry at St. Cloud was used for the basement walls, and Georgia marble for the superstructure and dome. Polished Mankato-Kasota stone, as well as a variety of marbles, form the interior walls. Granite columns quarried at Ortonville and Rockville stand beside the dome piers in the rotunda, and Pipestone rims the inner dome.

The interior features marble from around the world. The inlaid rotunda floor, designed by Gilbert, uses Numidian marble from southern Africa for the reddish points that radiate from a ring of dark blue marble from the state of Georgia. A 1905 review in the magazine *Craftsman* captures the elegance: "The great staircases of the Rotunda are resplendent with marbles—Hauteville and Echaillon from France, Skyros from Greece, and Old Convent Siena and Breche Violette from Italy.... The Senate Chamber is a square room finished in French Fleur de Peche marble, with its creamy ground and violet coloring." The Capitol merits a slow-paced and guided visit.

Capitol rotunda.

Capitol stair.

Capitol stair.

Exterior detail, Capitol.

19

Minnesota Judicial Center < 1918 >

25 Rev. Dr. Martin Luther King Jr. Boulevard

Judicial Center.

20

State Office Building < 1932 >

100 Rev. Dr. Martin Luther King Jr. Boulevard

Part of Cass Gilbert's original plan for the capitol complex included a mall, lined part of the way by government structures that would have extended from the Capitol to the Mississippi River. The plan was never carried through, but both location and style of this roughly matched pair, flanking the Capitol, give some idea of Gilbert's original design. Both buildings are rectangular granite structures with classical details in their columns and pediments. The judicial building (from 1917 to 1991 the Minnesota Historical Society building) was constructed of gray granite quarried at Sauk Rapids. Its main staircase and corridor floors are of Makato-Kasota dolostone and the walls of its vestibule and entrance are a dolostone from Frontenac, just south of Red Wing. The State Office Building is faced with granite quarried southwest of St. Cloud. Both are Clarence Johnston buildings: Senior designed the older one and Junior the younger.

21

Transportation Building < 1958 >

395 John Ireland Boulevard

Rockville granite panels cover the exterior. The building's design reflects soulless international architectural styles of the 1950s, when machine-made components had long since replaced handcrafted stone.

22

Veterans Service Building < 1954 >
20 West Twelfth Street

A new type of gray granite was added to the Capitol complex when Diamond Gray from Isle, Minnesota, was chosen for the exterior of the Veterans Service Building. The eight oval pillars are cut from Agate granite from Ortonville, Minnesota. The courtyard is paved with Crab Orchard flagstones, a quartzite from Crossville, Tennessee. Trap rock (basalt quarried near Dresser, Wisconsin) has been used for a retaining wall.

Veterans Service Building.

To critic Larry Millett, this is "the best post–World War II work of architecture on the mall and one of the city's first truly modern buildings of note. It has been criticized for standing square in the middle of Cass Gilbert's grand vista; however, by the time the building was planned, Gilbert's dream of a mighty boulevard marching south from the mall was long since dead."

23

Centennial Office Building < 1958 >
658 Cedar Street

The multicolored granite mosaic at the base of the Centennial Office Building contrasts with the panels of gray granite, metal, and glass that rise above it. This is the only hint of imagination in a building designed, apparently, to lament Minnesota's first century of statehood. Cass Gilbert would shudder at the dispiriting ugliness of this, the Transportation Building, and the Armory disfiguring his Capitol mall.

24

St. Paul National Guard Armory < 1961 >
600 Cedar Street

The exterior walls are Rockville white granite from Cold Spring, Minnesota.

25

Orville Freeman Building < 2005 >
625 North Robert

26

State Laboratory Building < 2005 >
601 North Robert

These two new state office buildings display a style and imagination that are a welcome relief from the stodginess of the Centennial and Transportation Buildings just a few blocks away. Both are made primarily of Mankato-Kasota stone, though of different shades (the warmer belonging to the Laboratory), and the Freeman Building has a granite base of Academy Black. Of Freeman Larry Millett writes, "With its bold play of vertical and horizontal elements and the spatial complexity of its façade, the Freeman Building has a Ralph Rapson-like swagger to it."

State Laboratory.

27

Harold Stassen Department of Revenue Building < 2005 >

600 North Robert Street

This stands across Robert Street (on the east side) from the two previously described buildings. The exterior walls are a composition material designed to resemble granite, but the only actual stone is Kasota, which is used for the window sills and trim.

One.

The garden on the west side of the building has nine sculptures by St. Paul artist Steven Woodward, representing the numbers 0 through 9. Each sculpture seems to be made from a different granite; the 0, Carnelian from Millbank, South Dakota; the 1, Rainbow Stone from Morton, Minnesota; the 2 Agate granite from Ortonville; the 3, Royal Sable from Millbank, South Dakota; the 4, Lake Superior Green from Isabella, Minnesota; the 5, Axalea, from Llano, Texas; the 6 and 9 (the same figure serves for both), Lac DuBonnet from a town of the same name in Manitoba; the 7, Lake Placid Blue from Jay, New York; the 8, Rockville White from Rockville, Minnesota. The figure representing two has the numbers back to back forming a heart shape—suggesting that the Department of Revenue has a heart. No doubt it does.

Two.

Together these three state government buildings mark a heartening improvement in attention to beauty, especially compared to the horrors of the 1950s and 1960s. Government buildings do not have to be ugly.

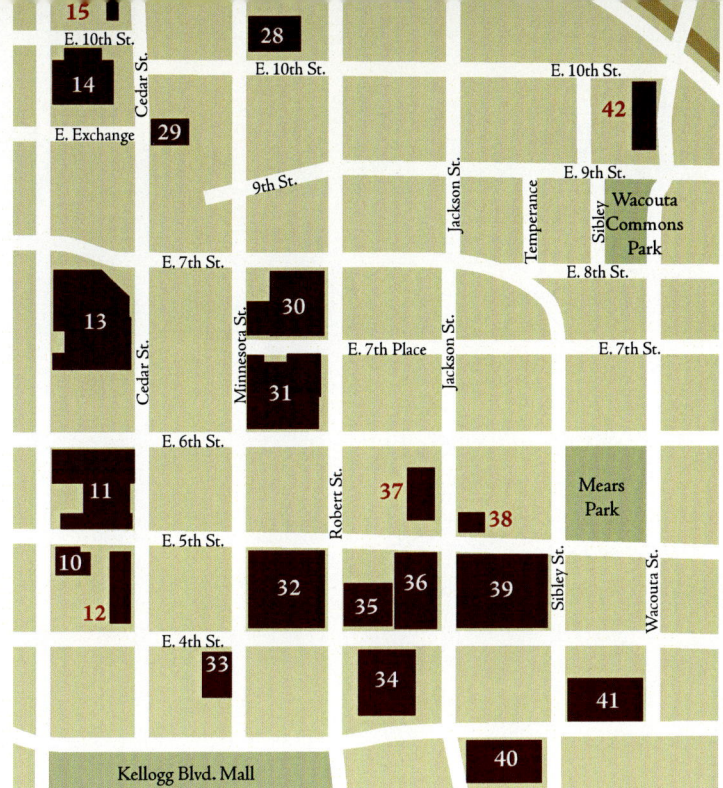

28. Public Safety Building
29. Central Presbyterian Church
30. Golden Rule Building
31. Securian Center, 401 Building
32. First National Bank
33. Minnesota Building
34. Warren E. Burger Federal Courts Building
35. Pioneer Building
36. Endicott Building
37. Agribank
38. Brooks (McColl) Building
39. Railroad and Bank Building
40. Main Post Office
41. Union Depot
42. First Baptist Church

Public Safety Building   < 1930 >

101 East Tenth Street.

Until recently this was the main St. Paul police station and fire department headquarters. It is slated to become part of a condominium complex. The neoclassical style, especially before the insertion of the disfiguring skyway, conveys a restrained authority appropriate to its erstwhile public safety mission. The walls and columns are a pale Hinckley sandstone; beneath them is a Minnesota granite.

29

Central Presbyterian Church < 1889 >
500 Cedar Street

Another massive pile of sandstone, the deep red-brown color courtesy of its high iron content. The stone comes from Bayfield, Wisconsin. The wise visitor steps inside to enjoy the fan-shaped sanctuary and the stained glass. This church is on the National Register of Historic Places.

Central Presbyterian Church.

Bayfield sandstone plaque.

30

Golden Rule Building < 1915 >
85 East Seventh Place

This is the last survivor of the great downtown department store buildings—Dayton's, Mannheimer's, The Emporium—this one, now converted to offices, retains the heft and elegance of the days when downtown merchants ruled retail commerce. The bases of the exterior walls are Carnelian granite from Milbank, South Dakota, and the main walls are limestone. Just below the cornices one can see large rectangular panels of Vermont Verde Antiqua, a marble-like (but not marble) stone from Rochester, Vermont.

A corner of the Golden Rule Building.

31

Securian Center, 401 Building < 2000 >

401 North Robert Street

The shorter and newer (by 19 years) of Securian's (formerly Minnesota Mutual) two headquarters towers uses three varieties of granite, Pioneer Red from India, Sunset Red from Texas, and Rockville White from Minnesota. The taller, duller tower is just gray granite.

Two shades of granite, Securian Center.

32

First National Bank < 1931 >

332 Minnesota Street

This is the tallest of St. Paul's Art Deco buildings, and, like so many of them, completed in the early days of the Great Depression. As is true for these others, most of the stone is limestone from Indiana. It was the city's tallest building until the 1980s. An early skyway connected this tower with its 1915 predecessor to the east. The long lower lobby, reachable from either Cedar or Robert streets, features Art Deco details and many yards of lovely, polished marble in three colors and varieties.

First Bank.

First Bank lobby.

33

Minnesota Building < 1930 >

42–48 East Fourth Street

This building features Indiana limestone above, with a base of Rainbow Stone. According to Larry Millett, the entrances on Fourth Street and Cedar Street are "delicately detailed" and "very urbane."

34

Warren E. Burger Federal Courts Building < 1969 >

316 North Robert Street

This Federal Courts Building replaced the Landmark Center as the local seat of federal justice in 1969. Its lower walls are of granite (trade name, Charcoal Black) that contrasts boldly with the white Vermont marble of the piers and upper walls. The stone is handsome, but, in the words of Garrison Keillor, the building is "a file cabinet with a marble veneer."

Warren Burger U.S. Courthouse.

35

Pioneer Building < 1889 >

336 North Robert Street

St. Paul's first skyscraper, the twelve-story Pioneer Building was erected in 1889 at the corner of Fourth and Robert Streets. Constructed of Rockville granite and red press brick, the building has an iron skeleton. Four more stores

Pioneer Building.

were added in 1912. Until 1915, this was the tallest building between Chicago and the West Coast. It was designed in the Romanesque Revival style by Chicago architect Solon S. Beman. The Pioneer Building has the traditional base-shaft-cornice pattern used in structures with load-bearing walls. A magnificent interior court extends from the second floor to the sixteenth. Cast iron pillars support balcony promenades and there is a wrought iron spiral staircase. Its elevator shafts are open.

A 1910 postcard view of the Pioneer Building. Photo courtesy of the Minnesota Historical Society.

36

Endicott Building   < 1889 >
141 East Fourth Street

Cass Gilbert and John Knox Taylor designed the building in the style of an Italian Renaissance palace. The color and shape of its rusticated sandstone base are repeated in the rectangular brick walls. Arches and lintels of carved Portage Entry Sandstone frame each window. The decorative touches in the entrance arch and central window in the elaborate cornice help make it a visually interesting building.

Larry Millett has called the Endicott, Pioneer, and Empire Building (next door to the Endicott and disfigured in the 1960s by stone

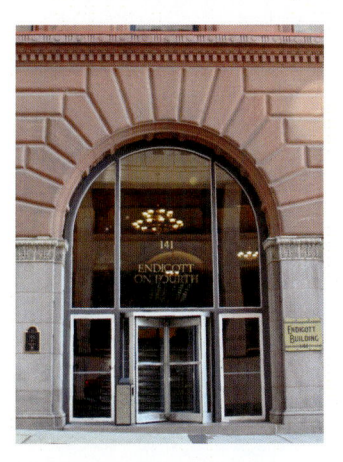

Endicott Building entrance.

Endicott Building cornice.

< 31 >

Endicott Arcade.

sheathing on the ground floor) "the finest ensemble of late nineteenth century office buildings in the Twin Cities." Just around the corner, on Fifth Street, the Endicott Arcade features a classical cornice and pediment, in fossiliferous limestone, unique downtown.

37

Agribank < 1967 >
375 Jackson Street

A file cabinet with a marble (white Vermont, from Dorset Mountain) veneer.

38

Brooks Building < 1890 >
366–368 Jackson Street

Formerly the Merchant's National Bank, the tiny lobby features eight varieties of marble. The exterior is Portage Entry Sandstone from Marquette, Michigan, with some St. Cloud granite. The architect was St. Paul's own Edward Bassford. The sign may say Brooks, but to St. Paulites this will always be the McColl Building.

Brooks (McColl) Building.

39

Railroad and Bank Building < 1916 >
180 East Fifth Street

For 57 years, until the opening of the IDS Building in Minneapolis in 1973, this was the Twin Cities' biggest office building. James J. Hill built it at the end of his life as the headquarters of his Great Northern and Northern Pacific railways. The heroic pillars, suitable for an empire builder, are made of Hinckley sandstone.

Railroad and Bank Building.

40

Main Post Office < 1934 >
180 East Kellogg Boulevard

Very restrained Art Deco on the outside, but inside there are some nice details. Visit it soon, before it comes down. Its future, like its design, is grim. The stone is Mankato-Kasota.

Main Post Office.

41

Union Depot < 1923 >
214 East Fourth Street

This one opened in 1923, replacing a previous Union Depot destroyed by fire, just in time for the beginning of the decline of passenger rail traffic in the United States. Still, it had a useful life of fifty years. For nearly forty years now it has been searching for a reason to stick around, and the search goes on. Since the last passenger train pulled out in 1971, the

< 33 >

space has been used for offices, restaurants, and, soon, condominiums. Hope remains for its return to rail service.

The depot's front is dour, massive, broad-shouldered—James J. Hill would have loved it had he lived long enough to see it. The style is Beaux-Arts, the columns Doric, the stone Indiana limestone (like City Hall). Step inside for a view of the vast main hall, and try to imagine it busy with

Union Depot pillar.

people, leaving, arriving, waiting, or working. The interior walls are covered with panels of Kasota dolostone, pink Tennessee marble, and gray Missouri marble.

42

First Baptist Church < 1875 >
499 Wacouta Street

Erected in 1875 (making it the oldest building in Lowertown), this Gothic Revival building is unique to downtown for its bush-hammered finish, a heavily contoured surface quite in contrast to the smooth finish of the many other Mankato-Kasota stone buildings in the city.

The interior merits a visit. The dark wooden beams are very fine, the organ played at the church's opening, and in the southeast corner one can

Pointed arch, First Baptist Church.

easily see the settling that required the original steeple to be removed in 1945. This is a building with roots. Harriet Bishop, St. Paul's storied and celebrated (Harriet Island) first public school teacher, worshipped here.

< 34 >

A BRIEF GEOLOGICAL HISTORY OF MINNESOTA

The stone used in the construction of St. Paul buildings came from a geological past that has a history of surprising diversity. The land that is now Minnesota was there billions of years ago when the earth was young. Its most ancient rocks are about 3.5 billion years old and are found in the Minnesota River valley near Morton and Montevideo. They are a granitic gneiss formed from even earlier sediments that had been changed by heat and pressure generated within the earth.

Other granites—those that are found at St. Cloud and Isle—are younger, 1.8 to 1.5 billion years old. Granite quarries today work the roots of what were once deep chambers of plutonic rock, the result of a period of recurring igneous activity. Minnesota iron ores and Sioux quartzite, with its interbedded Pipestone layer, were originally deposited more than a billion years ago in the epicontinental seas.

Morton gneiss. Photo by the Minnesota Geological Survey.

Volcanic activity again shaped the landscape about a billion years ago. The Chengwatana Volcanics, source of the trap rock at Dresser, Wisconsin, poured out as molten rock, or lava. Later sediments accumulated to great thicknesses in another epicontinental sea. These sediments formed the brown, red, and buff sandstone of the Hinckley formation and the Bayfield group, and were the brownstones used so extensively during the 1870s through the 1890s as dimension stone.

Sediment deposited 500 million years ago formed the Late Cambrian rocks that can be seen in outcroppings on the St. Croix River bluffs. This Croixian series of sandstones, shales, and limestones reflects the history of the seas that transgressed and regressed over the region.

Recurring deposits continued during the Ordovician Period. The Oneota dolostone, which is about 450 million years old, carries evidence of marine life. Platteville sediments are interbedded with shale partings—weakening the limestone blocks for construction use. Decorah Shale, a source of clay for brick, came from an environment teeming with life and contains fossil sponges, bryozoans, bivalves, cephalopods, crinoids, brachiopods, and trilobites.

Covering these bedrocks are the more recent glacial deposits which are older than 10,000 years. There is evidence that at least four glaciations occurred in North America. The last, or Wisconsin, glaciation can be seen in the tills of the Superior and Des Moines lobes that shaped the Minnesota and Mississippi river valleys. Their gravel deposits surround the Twin Cities and are used in modern construction.

THE QUARRIES

Sandstone-Banning. Quarrying of building stone is an ancient and enduring art. Drills, plugs, feathers, wedges, chisels, and hammers were used to build the Egyptian pyramids, the Great Wall of China, and the Roman Forum, and were still the basic equipment brought to America by the English quarrymen who came to work in the Kettle River quarries.

The Banning quarry, begun in the 1870s, and the Sandstone quarry, begun in 1885, supplied stone for hundreds of miles of paving and curbing in St. Paul and Minneapolis. The stone also has been used for railroad bridges, for the wing dam at St. Anthony Falls, for riprap along the railroad beds, for the foundation of Minneapolis City Hall, and for the dome piers of the state capitol. It can be seen as the cream-colored cross-bedded sandstone used in the base of the Burlington Northern office building in St. Paul. Because the sand grains are held together by silica cement, the stone is both durable and impervious to dampness or frost.

Today the Sandstone quarry has been developed into Robinson City Park and only remnants of the quarrying industry remain among a growth of birch and aspens. Upstream, near the dalles of the Kettle River, the old Banning quarry has become part of Banning State Park.

Bayfield-Houghton-Washburn-Apostle Islands. Bayfield, Wisconsin, famous for its scenery and its sandstone quarries, was founded in 1856 by Henry M. Rice, one of St. Paul's illustrious pioneers. Bayfield sandstone was a choice building material widely used in St. Paul in the 1870s, 1880s, and 1890s. Central Presbyterian Church is an excellent example of the use of No. 1 Lake Superior brownstone from Captain R.D. Pike's quarry at Bayfield. A medium-to-coarse-grained sandstone, it has a small amount of iron oxide that gives it a reddish color.

The Prentice Brownstone Company near Houghton Point, south of Bayfield, began quarry operations before 1857. In

A 1943 postcard view of a granite quarry in St. Cloud. Photo courtesy of the Minnesota Historical Society.

1892 the company launched an ambitious project: the carving of a great sandstone monolith for the World's Columbian Exposition in Chicago. Larger than Cleopatra's Needle in New York's Central Park, the monolith was intended to dramatize the dominance of sandstone as a building material. The Panic of 1893 ended the work and the monolith never was removed from the quarry. Changes in the construction industry soon ended the use of expensive heavy masonry.

Babcock and Smith, with headquarters at Kasota, Minnesota, and Washburn, Wisconsin, operated the Washburn Stone Company near Houghton. The quarry opened in 1885 but is abandoned today.

The Apostle Islands' sandstone quarry industry began on Bass Island and produced brownstone, which was widely used to rebuild Chicago after the fire of 1871. The Basswood (Bass), Wilson (Hermit), and Stockton (Presque Isle) Island quarries now are part of Apostle Islands National Lakeshore Park.

Marquette, Michigan A particularly desirable building stone called *Portage Entry* was quarried from the Jacobsville formation

Jacobsville sandstone, Upper Peninsula of Michigan, St. Paul Building.

near Marquette, Michigan. This sandstone was known for its fine-grained and even texture, and its uniform color. The quarries were located in the Keweenaw Peninsula of Upper Michigan and the stone could be transported easily by boat and rail to Midwest cities.

Kasota and Mankato. In 1851, as partial payment for delivering a consignment of mules to Fort Snelling, Joseph W. Babcock was given a tract of wooded government land at Kasota, Minnesota. Here he set up headquarters for his St. Paul to Sioux City mail service. And because the Minnesota River at this point had conveniently eroded the soil, leaving exposed rock ledges containing 20 to 60 feet of Oneota dolostone, Babcock also opened a quarry. Similar quarries were opened in the Minnesota River valley between Kasota and Mankato, and they have produced a fine building material, the pink, cream, buff, and gray Kasota and Mankato dolostone.

Bush-hammered Portage Entry sandstone, Endicott Building.

Because it is durable, Oneota dolostone lends itself to modern building techniques. It can be sawed into thin slabs; it resists weathering; and it is easy to maintain. The Oneota formation is part of the Prairie du Chien group, an Ordovician carbonate rock that underlies much of southeastern Minnesota. Where it outcrops along river bluffs, it has been quarried for dimension stone. Old quarries at Frontenac and Red Wing are now state and city parks.

Winona. The Biesanz Stone Company began quarrying at Winona, Minnesota, not long after the Civil War. The company has always produced just one thing, Winona travertine, its trade name for Oneota dolomite, also known as Mankato-Kasota stone. The name "travertine" (which ultimately derives from the Tiber—*Tevere* in Italian—River of Rome) refers to its pocked and gouged-appearing texture. This

History Center floor—Winona Travertine, Rockville white granite, and white Georgia marble.

particular dolomite is hard enough to take a polish, and can be made to resemble marble, as it does, for example, in the floors of the Minnesota History Center. The quarry on Stockton Hill at Winona, opened in 1913, contains the deposits of the inland Ordovician seas that covered the area about 450 million years ago.

The St. Cloud Area. Diamond Pink, St. Cloud Red, Opalescent, Richmond Green, Dark Pearl, Charcoal Black, Diamond Gray, and Reformatory Gray are some of the trade names for the colors available in granite from the St. Cloud-Isle area. The range of tints is broad because of the variety of minerals present in these durable rocks. The St. Cloud gray stone is the oldest igneous rock in the area and is dated at 1.8 billion years.

Rockville granite, detail, St. Paul Cathedral

Granite quarrying in Minnesota began around 1868 near Sauk Rapids, and over the years many small operations have come and gone. In 1889 a youngish Scotsman, Henry Alexander, and some stoneworking partners formed their own operation, the Rockville Granite Company. This one stuck around. Renamed Cold Spring Granite Company in 1924, it is now one of the nation's largest granite producers, with over thirty quarries, five fabricating plants, and more than a thousand employees in several states and Canada. The Alexander family has stayed in the company all along. At the end of 2007, a grandson of the founder, Patrick Alexander, is chairman of the board.

GEOLOGICAL TERMS

Basalt Fine-grained, dark-colored volcanic rock. It is "extrusive," meaning it poured out onto the surface of the earth, where it cooled rapidly. The rapid cooling prevented the formation of large crystals; hence the fine grain.

Cambrian Time period dating approximately 500 million years ago or rock formations from this period.

Cross-bedded Series of laminations found inclined to the main bedding plane of granular sediments.

Dolomite A mineral name for calcium magnesium carbonate. It's a sedimentary rock similar to limestone, which is mainly calcium carbonate. The difference is the dose of magnesium.

Dolostone Another word for dolomite, a stone usually referred to here as Mankato-Kasota stone.

Epicontinental seas Shallow portions of ocean that extend into the interior of a continent.

Gabbro An intrusive igneous rock, dark in color, chemically equivalent to basalt but coarser in texture.

Gneiss A banded, coarse-grained metamorphic rock, often composed of granite and allied rocks. The term thus refers to its appearance rather than to its chemical or mineral composition.

Morton gneiss, detail, Qwest Buildings.

Granite A coarse-grained igneous rock. It is "intrusive," meaning that from its melted form it cooled beneath the surface of the earth, hence slowly. The slow cooling permitted the formation of the crystals that one can readily see in all granite.

Hinckley Sandstone A sandstone quarried near Hinckley, Minnesota.

Igneous Rock crystallized from a molten or partially molten state.

Interbedded Occurring between beds or layers of rock.

Kasota stone Dolomite from the Mankato-Kasota area of the Minnesota River valley. Also known as Oneota dolostone and Mankato-Kasota stone.

Kettle River variegated Trade name for Hinckley sandstone.

Lava flow Fluid or viscous lava flowing from a fissure or volcanic cone; also, the hardened result of that action.

Limestone Bedded sedimentary rock consisting chiefly of calcium carbonate.

Magma Molten material within the earth from which igneous rocks are derived.

Marble A metamorphic rock composed of calcium carbonate and dolomite. It is essentially limestone or dolomite that has been subjected to extreme heat.

Onyx A form of quartz made up of tiny crystals. Chemically, quartz is silicon dioxide, the second most common mineral on earth.

Indiana limestone, detail, City Hall/Courthouse.

Ordovician The time period immediately following the Cambrian; rock formations dating from this time period.

Outcrop Bedrock projecting through the cover of soil.

Pipestone Catlinite, a shale layer in Sioux Quartzite that was used by Dakota people for carving, often of ceremonial pipes, hence the name.

Platteville limestone A variety of brittle limestone characterized by conspicuous layering and rough texture that takes its name from Platteville, Wisconsin.

Plug and feathers Wedges used in quarry work.

Plutonic rock Igneous, intrusive rock that crystallized from a magma below the surface of the earth. Granite is an example.

Platteville limestone, Assumption Church.

Precambrian Rocks formed before the Cambrian time period; time period of the same name.

Rainbow Stone A trade name for Morton gneiss, a metamorphic rock quarried near Morton, Minnesota.

Sandstone Cemented or compacted sediment composed predominantly of quartz grains.

Sediment Solid material settled from suspension in a liquid.

Shale Layer sediment in which the particles are mainly clay size.

Till Unsorted sediments carried or deposited by a glacier.

Trap rock Dark colored dike and flow rocks, chiefly basalt and diabase.

Travertine Calcium carbonate of light color; compact, banded varieties when polished are called onyx marble. It is not marble, but rather a variety of limestone. Its characteristic finish is pocked and gouged.